I0154190

Amory Howe Bradford

Old Wine; New Bottles

Some Elemental Doctrines in Modern Form

Amory Howe Bradford

Old Wine; New Bottles
Some Elemental Doctrines in Modern Form

ISBN/EAN: 9783337329969

Printed in Europe, USA, Canada, Australia, Japan

Cover: Foto ©berggeist007 / pixelio.de

More available books at **www.hansebooks.com**

OLD WINE: NEW BOTTLES

ental Doctrines in
rn Form

DFORD, D.D.

D LIFE," ETC

RD & HULBERT
1892

COPYRIGHT IN 1892
BY AMORY H? BRADFORD

CONTENTS.

I.

THE LIVING GOD.

" THE being of God has not its foundations in the life of humanity : but humanity has its foundations in the life of God."—MULFORD'S *Republic of God.*

> " To Thee we rise, in Thee we rest,
>   We stay at home, we go in quest,
>   Still Thou art our abode.
>   The rapture swells, the wonder grows,
>   As full on us new life still flows
>   From our unchanging God."
>
> T. H. GILL.

> "Eternal Light ! Eternal Light !
>   How pure the soul must be,
>   When, placed within thy searching sight,
>   It shrinks not, but with calm delight
>   Can live and look on Thee ! "
>
> THOMAS BINNEY.

" Of the divine life itself there are no differing dispensations . . . Moreover, the divine attitude toward man was the same before the coming of our Lord as afterward. The dispensation which we call Christian, while it is the special, is not the only dispensation of grace. In its largest meaning the Christian dispensation is not limited to any time.—*God in His World.*

" And Simon Peter answered and said : Thou art the Christ, the Son of the Living God."—MATTHEW xvi. 16.

I.

THE LIVING GOD.

BUT why the *living* God? Because, in the minds of many, God is practically dead: He is a name, or a force, or a Being who once was near to the earth, but who has departed, and has no more interest or care for humanity; a Being who walked with the patriarchs, talked with the prophets, did marvelous works in apostolic times, but who has now retired into awful and unapproachable mystery, if not into inaction and death. There are two classes to whom God is practically dead—materialists, who believe in no spirit pervading the universe, who see nothing but matter and force working in accordance with unintelligent laws toward an end which must be death; and those who have a clear and strong faith in the God of the Hebrews, the God who manifested himself in Jesus Christ, who inspired the Apostles, and who continued in the world until the Bible was completed—and then disappeared forever. They do not say that he has

gone forever, but their theories concerning religion imply that so far as he is related to the
universe, and especially to men, he might as well
have no existence. Beside these cold and cheerless speculations—the first of which results from
undue emphasis upon phenomena, and not enough
upon spirit; and the second from a tendency
in human nature to exalt the past, which is
found not only in Christian nations but also
among the heathen, and which is the same in
reality as Ancestral Worship in China—I place
the teaching of Holy Scripture, which thrills with
inspiration, hope and power. Peter said, "Thou
art the Christ, the Son of the living God," and
the emphasis is no more on the first clause than
on the second.

The phrase, and the idea, "living God" is
common, both in the Old Testament and New.
The Patriarchs believed in a Being near and real;
he communicated his will; he talked with them
as they walked by the way; in the silence of the
night, beneath the near and tender heavens, he
uttered his commands; in dreams his messengers
seemed to descend and ascend on ladders of
light; he manifested himself in the beautiful
order of nature, and spoke with audible voice
unto those who had ears to hear. Moses com-
muned with him on the Mount, and received
from him laws for the people; Isaiah saw his
glory as it filled the Temple; Elijah heard his
voice, and was ministered to by his messengers

at the brook Cherith ; Daniel was delivered by
his power from the fiery furnace and the den of
lions. One phrase condenses the faith of the
Hebrews—*They believed in the living God.* When
we turn from the Old Testament to the New
the same vivid consciousness continues. The
Master lived in the realization of the Divine
presence and power : he talked with his Father
and held communion with him on the still mountains, in the quiet of Galilean midnights. The
Apostles in the same faith continued his ministry. Wherever he went Paul carried the message
—Ye are the children of the living God. He
told the Corinthians that they were the temples
of the living God ; he exhorted the Thessalonians
to turn from their idols and serve the living
God ; to Timothy he spoke of the Church of the
living God, and to the Athenians declared that
even idolatrous Greeks lived, moved, and had
their being in him. The author of the Epistle to
the Hebrews said that it was a fearful thing to fall
into the hands of the living God ; and John, last
and noblest of the Apostolic band, whose spiritu-
ality seemed to have acquired a finer sensibility,
spoke of the time when all will come to Mount
Zion, to the city of the living God.

Throughout, the Bible palpitates with life. It
is the Book of life ; its teachings are conveyed
in terms of life. The God of the Hebrews was
not an impersonal power, distant and difficult of
approach, but near and constantly accessible.

When he manifested himself in his supreme revelation it was not in a book about which men could wrangle, but in a life which was its own best evidence. When the Master himself disappeared, he promised that the same living ministry should be continued, and that the Comforter should come, the Spirit of Truth who should abide forever. In the New Testament God is always represented as near, efficient and unchangeable, "the same yesterday, to-day and forever"; not "a marble Deity upon a marble throne," but a Father, a Friend, a Helper; not a cheerless power, but a free Spirit. In him, men live and move and have their being. He is near them, even in their hearts, and yet he fills all spaces, for if we take the wings of the morning and go to the uttermost parts of the sea, he is there, and if we make our bed in the depths of the under-world, he is there. Three thoughts are always prominent in the Scriptures:—God is living; God is a Spirit; God is everywhere.

Philosophy and theology have been steadily moving toward a fuller appreciation of these central and governing truths. At first God was regarded as transcendent; that is, a Being beyond the blue of the sky and the shining of the stars, or at least beyond the possibility of approach. Then men began to ask how such a Being could have any relations with humanity; and thus gradually, opinions have crystallized in the belief that God is not throned in splendor outside the walls of the universe, sending his messages down to humanity, as a king sends messages to his far-away subjects, but that he is spiritually every-where. Do not be troubled by that expression. We know not what spirit is, and yet are conscious that we are spirits. Each day we are humbled by the greatness of our own ideas. Imagination can traverse the universe in an instant; one mo-ment we are here, and the next on the remotest of the fixed stars. Thought flies more swiftly than light; a hundred million miles in a second is no unusual speed. Imagination is unlimited. You may shut a man's body in a room, but his imagination and his memory will leap all barriers and the one will go to the ends of the earth and back along the track of years, while the other will soar through the spaces, limitless and free, as if there were no body. *The Pilgrim's Prog-ress* was probably conceived by John Bunyan while he was in Bedford Jail, but bolts and bars of iron could not prevent him from climbing the Delectable Mountains and walking the streets of the Celestial City.

Just what is meant by the phrases, "God is a spirit," and " God is everywhere," we do not know. There is nothing contradictory in them, but the thought is too great for us to compre-hend. We do not know what space is, or what eternity is, or even much about what the ocean is, although we talk of these things as if we knew. Modern theology is characterized by a

doctrine of the Divine Immanence is not new,
but the old teaching of the Omnipresence made
to mean something.   And that so-called "New
Theology," which so many fear will overturn
faith in religion, really is an incalculable blessing
because it is a return to this fundamental truth.

"In the beginning God created."   In the "beginning:"—imagination is wearied in trying to
exhaust that word.   In the beginning, God: at
the end, the City of the Living God.   Thus we
have in the Scriptures the living God at the
beginning of all things and at the. end of all
things.   Between these two poles the Patriarchs,
the Prophets, the Psalmist, the Apostles bore
vital witness to the existence and the influence
with men of the living God.   There is no uncer-
tain strain in all that music.   Behind all things
is the living God, controlling forces, inspiring
processes, the author of life; manifest in an
audible voice, in a flaming bush, in a vision of
the night, in a great historic movement, at last in
a human form—a unique personality: and from
first to last permeating all things, living in the
progress of humanity, reflecting his glory in the
splendor of the skies, revealing his watchful care
in the clothing he fashions for the birds, and the
provision which he makes for man.   We are not
in the hands of blind Force, or of One so far
away that he cannot understand, and does not
care for us: but life and history are guided by
return to faith in the omnipresence of God.   The

the same intelligence which presided at the Creation, and the same love which bled upon the cross; and we, howsoever humble, may walk with God and receive inspirations from him in the same way and to the same extent as patriarchs and prophets, apostles and martyrs.

The doctrine of the Living God implies that all things in the past have been in his hands, and no man and no nation have ever been separate from him. His providence was not limited to the Hebrews: it was conspicuous in their history and he chose them for a special work; but he has had other purposes and other lessons to teach mankind than those which came through Israel. He was as near to Babylon and Damascus as to Jerusalem and Bethlehem: the same hand that smoothed the slopes of Calvary and piled the snows on Lebanon made a passage through the mountains for the Colorado, and lifted toward the sun the glittering glory of Shasta. The living God could not be a little God. Since all things have been in the hands of the living God, love, justice and truth have been at the heart of things in all the past. Whether humanity started in a state of innocence but of imperfection, or whether there was in the past some golden age from which there has been a fall, is of little practical importance. Whatever the theory of the development of man, God has never been absent, and his care for the children whom he has made has never wearied.

"Whatever is, is right,"—a more untrue word
was never spoken. Are vice and crime right?
Is the Devil a saint? Many things not "right"
are *best* for localities or conditions, but they must
be gotten rid of as soon as they have served
their purpose ; they are only means by which
the right and truth are hastened. The old sac-
rificial and ceremonial system was not right, and
the prophets were hot against it, but it was best
under the circumstances. Many abuses have ex-
isted, from which men argue that there can be no
justice or love ; but they forget that those abuses
were only conditions through which, in a natural
and living way, a higher and more perfect state
has been realized. A bandaged arm is not right,
but it is a condition by which a broken arm is to
be made right. Why was the coming of the
Master so long delayed? Why were men, fitted
for a religion of light and love, so long permitted
only the light of nature? We are pointed to the
mysteries of Egypt and Greece, to the imperfect
moral teachings of the sages and seers of other
faiths, and asked, If there was a God how could
he allow such things? The Master said to his
Mother, "Mine hour is not yet come ;" and his
hour had not come a thousand years before the
Advent, or five hundred years, or one year before
it. But it cannot be argued that therefore God
had no interest in humanity, and that the millions
who peopled Asia, Africa and Europe, and who
perhaps in earlier times roamed between the

Atlantic and Pacific, were absent from the watch and care of the Almighty. If he was living he was living everywhere. Either there was no God, or there was a Being unworthy of the name, or else we must believe what the Scripture says, that "he has not left himself without a witness among any people." For myself, I rejoice to believe that the same love which was manifested on Calvary has been manifested, according to the ability of men to receive it, in all the ages of the world's history, and among all sorts and conditions of men. There have been ruder times and greater darkness, but those who walked in darkness did not thereby cease to be God's children : and it is probable that in many lands those who have tried to do right even if they stood alone, have been taken from lions' dens like Daniel, and like Jacob have seen angels coming to them out of heaven. The living God has always and everywhere been the loving God, and his hand has never once been off the wheel of change.

All things *are* in the hands of the living God ; therefore the Christian doctrine of Providence is a necessary and beautiful reality. It is difficult to understand how one Being can see and know all who live ; how he can hear their prayers, and care for their necessities ; but difficulty of comprehension is no indication of unreality. We speak the word " God," but no one ever yet compassed its meaning. We talk about the Creator, but who has a clear idea of what "Creator" signifies? We talk of the universe, but who knows much about the universe? Day by day science reaches farther into the spaces ; some new star flashes its light ; and who can tell how many yet are undiscovered? Before infinite space we must bow with humiliation and awe. No man has fathomed it, and the seen is to the unseen as a single leaf to all the autumn forests ; as one grain of sand to all the sand of the seashore ; as one ray of light to the splendor which fills immensity. We comprehend few of the things by which we are surrounded, and as little appreciate the reality of Divine Providence. The Master said that He who clothed the lilies and fed the sparrows would much more care for the children of men : and a Being who has personal interest in all the birds, and all the flowers, and all humanity, is beyond the utmost stretch of imagination. Yet, in the midst of the storm and stress, when it seems as if infinite darkness were enshrouding all things, how sweet and glad to realize—as realize we may— that we rest in love fathomless as the spaces and endless as the ages! Infinite intelligence and infinite love were not the exclusive property of some far past age, or of a narrow and conceited people ; they belong to all peoples and all times. I cannot tell you how it is, but I am sure that no single child walks the streets whose name is not known in the heavens ; not a burden presses upon you in business or in your homes that does not

also rest upon the heart of God; not one life is
without purpose, and not one soul " walks with
aimless feet." Providence is a reality.

But, also, Inspiration is a reality. Inspiration
is the Divine Spirit speaking to the human, and
wherever the heart is pure and the mind open,
that Spirit comes as naturally, and inevitably, as
light into a house when the windows are opened.
It was a great privilege for Enoch, Moses and
Elijah to walk with God, and to hear him speaking
in a way which they recognized: but we are in
the same humanity as Enoch, Moses and Elijah.
If their hearts had been less pure, and their ears
more dull, they would have heard no accents of
the Divine voice ; and if our hearts are equally pure
and equally open, the same Being who spoke to
them will thrill us with the message which it is need-
ful for us to hear. Some imagine that the days of
miracles are past, and that the day of inspiration
is ended, but that is to say that God is dead, or
that humanity has reached His altitude. Rather
let us believe in continuous miracle and continuous inspiration ; that men are as truly inspired
to-day as ever they were ; that as hearts become
purer their inspirations will be more frequent and
constant. " The pure in heart shall see God "
is not limited to one time. There is no need of
another Bible, and none are now inspired to write
Bibles; but there is need of clearer light, and
wider knowledge in many spheres, and when our
Master went away he promised the Spirit of

Truth, who would guide into all truth and show things to come.

All the past has been in the hands of the living God, and the living God must be the loving God, for there can be no God without love. He did not cease to love and sacrifice when the cross was raised on Calvary. All things in the present are in the same hands, and therefore where there are pure spirits the Divine Spirit reveals himself to them as to prophets and apostles. Some think they cannot believe in providence and inspiration. Let us rather say we cannot disbelieve in them. They are necessary to the idea of God.

A little child was lost the other day in the streets of New York, and after wandering long was picked up and cared for by some kindly people. Unable to tell where he lived, or by what way he had traveled, I do not know whether even yet he has found his home. But if that little child could have wandered away and have been, for one moment, out of the sight and care of God, I do not see how there could be a God. Along the track of the old emigrant roads, across the great Western plains, here and there are piles of stone, and above them a simple cross of wood, weather-beaten and decaying. When travel was by wagon, emigrants moved along those roads, and now and then a man, or woman, or child died, was buried by the way, and the procession moved on. Who lies beneath those rude cairns no one can tell, but if one of those persons is

unknown to God I do not see how there can be
a God.   It is not difficult to believe in providence :
rather, it is impossible not to believe in it and
retain our faith in God.

The future is in the hands of the living God ;
he is the same yesterday, to-day and forever.
He was manifest at the Advent ; he was manifest when the Master spoke to the woman of
Samaria and told her all her secret life : he was
manifest when Jesus took little children in his
arms and blessed them ; he was manifest when
Jesus wept at the graveside in Bethany ; he was
manifest when Jesus hung upon the cross and
prayed for those who crucified him, " Father, forgive them, for they know not what they do."
" Yesterday, to-day and forever " !   All the past
on the heart and in the hands of him who offered
that prayer.   All the present on the same heart
and in the same hands.   All the coming days on
the same heart and in the same hands.   We
know not what may be the surprises of the
future.   Judged by the past, great and marvelous
will be the ways of the King of Saints.   New
discoveries, of which now we cannot dream, will
break upon the world.   The veil between the
physical and the spiritual will grow thinner, until
perhaps through it men will almost be able to
see.   The unexplored spaces will speak their
secrets into the ears of those who listen.   Conditions which now are vicious, and circumstances
which work ruin, will be transformed, and a new

and more healthful stock breed a race with purer
affections and loftier aspirations.   The future is
with God, and the Judge of all the earth will do
right.   Whatever is symbolized by the majestic
descriptions of the Judgment will come to its
realization; but the judgment-seat itself will be
occupied by Him who hung upon the cross, and
no harm and no injustice, and nothing at enmity
with love will come to one human being.

This is the faith of the Christian.   The heart
of our creed is not what we believe about the
Hebrews or the heathen, but what we believe
about God; and we believe that he was revealed
in Jesus Christ to take away the sin of the
world; that in all the past his plans of blessing
have been maturing; that in the present those
same plans are moving toward their consumma-
tion, and that sometime he will see of the travail
of his soul and be satisfied.   Things are moving
upward; the process never ceases, and God is
never absent.   He is in the struggle, submitting
to limitation, manifesting himself both in defeat
and in victory, and by defeat and victory, sorrow
and joy, hastening the day in which there will be
no more darkness and no more pain.   Nothing is
by chance; nothing is left to fate; all things are
in the hands of love.   Your child, your husband,
your wife, your friend, is in the grasp of a terrible
disease and your heart breaks; but God is not
dead, and he loves with a love transcending
yours, as the heavens are above the earth.   Your

child is getting into evil ways, and your heart
agonizes for him; but God is not blind, and his
love transcends your love, and his patience your
patience, as the heavens are higher than the
earth.

Here comes a question which possibly some
will be foolish enough to ask: "If all is moving
toward good, and all will work for good, what
difference does it make what we do? Are we
not doing his will whether we are righteous, or
whether we are wicked? If we do wrong is it
our fault?" Now and then such questions are
raised : what shall be answered? His purposes
for good will go on with us if we will, but without us if we resist. The mystery of freedom is
great and awful. It cannot be explained, but this
we know : the providence of God always works
toward harmony with wisdom, truth and love.
If any one follows evil he puts himself out of
harmony with the nature of things and the will
of God. Whether he can stay outside that will
forever, or whether by discipline and punishment
he will be compelled to come to himself, we
need not inquire. The outlook in either case is
terrible enough. It is never God's purpose that
any man should sin ; it is his will that all should
be saved. But, whether men will go with it or
not, his purpose of blessing will move on. A
steamer will leave the port of New York to-
morrow. You can go with it, or you can remain,
but the steamer will not be anchored to your

whims. And the plan of God, by which he is causing all things to work together for good, will be accomplished, and all may go with it, have part in it, and help to make its movement the swifter; they may have fellowship with the great and good of the ages; they may march in the same ranks with prophets and apostles, with seers and sages, with missionaries and martyrs— with those who have died rather than deny the truth; they may hasten the victory, and when it is complete, rejoice in the consciousness that they have helped a little toward hastening the glad new day. But a baby's hand might as well attempt to prevent the sun from rising as for any to imagine that they can do aught to prevent the realization of His purposes whose love is everlasting and whose power is resistless.

This, then, is the conclusion we have reached. We cannot grasp it, we cannot comprehend it, we cannot see it with the mortal eye, but we, and all who live, are in the hands of measureless, fathomless, omnipotent and unending Love. All the past is his, and all the future is his, also. We live and move and have our being in the Living God; and he says: "Whosoever will, let him come, and take of the water of life freely."

II.

THE HOLY TRINITY.

"There is only one God. The testimony of Scripture here absolutely accords with the witness of our conscience, and with the obvious unity of the universe in all its provinces and successions."—A. A. HODGE.

"God exceeds our measure, and must, until either He becomes less than infinite or we more than finite. If we can apprehend Him so as to be clear of distraction, and of terms that are absolutely cross to faith itself, it is all that can be hoped."—HORACE BUSHNELL.

> "For the loving worm within its clod
> Were diviner than a loveless God
> Amid his worlds, I will dare to say."
>
> ROBERT BROWNING.

> "So to our mortal eyes subdued,
> Flesh veiled but not concealed,
> We know in Thee the Fatherhood
> And heart of God revealed."
>
> JOHN G. WHITTIER.

"Go ye therefore, and teach all nations, baptizing them in the name of the Father, and of the Son, and of the Holy Ghost."—MATTHEW xxviii. 19.

II.

THE HOLY TRINITY.

THE doctrine which fifty years ago was most in the thought of American Christians, and which once divided the Church, is now seldom mentioned. It could hardly be otherwise. The subjects which interest to-day are practical rather than speculative; thought reaches down to fundamental questions, and is occupied with inquiries concerning the existence of God and the spiritual and immortal nature of man. Yet it is not to be supposed, because men talk less about the Trinity, that it has been relegated to the museum of theological antiquities. Its essential elements were never so firmly held or so widely accepted; but this and all other religious truths are approached in a better spirit than formerly. Those who without a respectable knowledge of their own language, and much less of the languages in which the Bible was written, presume to talk authoritatively about things in infinity and eternity, have fortunately almost disappeared.   Doctrines which never can be fully comprehended are not proper subjects for dogmatic statement, and least of all for controversy. Moreover, clear ideas about such subjects are now seldom required for church membership. Men are beginning to realize that as a violet may grow and absorb colors from the light without understanding its chemistry, so a spirit may grow in likeness to God without comprehending his nature.

The doctrine of the Trinity, then, is relatively out of sight, for two reasons: first, the progress of science has forced an entirely different class of subjects into prominence ; and second, Christian men appreciate that, however profound and practical the subject may be for meditation, it can never, without irreverence, be made a subject of debate.   The clouds of the old battle have blown away, and now it is seen that the combatants were fighting to the death for what in the nature of things could never be settled.   Each contended for a half-truth.   The Unitarians were strenuous for the unity of God, and the Trinitarians for his trinity.   Each side was true in what it affirmed, and largely in error in what it denied.

Laying aside further reference to the past, let us examine the doctrine of the Trinity from a Biblical and experimental standpoint.   I have no interest in the subject as mere speculation. As a theological conundrum it is entitled to no more consideration than a child's riddle.   If it

has no mission in the world except to be written
in creeds for the purpose of keeping Unitarians
out of so-called " Orthodox " churches the
quicker it disappears the better for spiritual religion.   Precisely because this doctrine as taught
by our Lord seems to me a condensation of the
Gospel, and an answer to the intuitions of the
reason and moral sense, I venture to speak to
you about it to-day.

Is the doctrine of the Trinity in the Bible?
The first feeling on turning to the Bible for the
purpose of finding this doctrine is one of amaze-
ment.   It is nowhere directly stated.   Let no
one be surprised, for the Scriptures do not teach
the existence of God.   They declare " The fool
hath said there is no God," but they do not
teach His existence ; they presume it from beginning to end, and without it would be meaningless.   In the same way they presume the Trinity,
and without it would be meaningless.   Our Lord
when his work on the earth was finished, when
he must have been overwhelmed with the consciousness that he would speak in the flesh no
more ; when he knew that his words would have
the authority of a final utterance ; when he was
giving his last directions concerning the advancement of his Kingdom, made the most exact and
emphatic statement of the doctrine that it has
ever had.   His disciples were soon to start with
their message around the world, and he was
about to leave them forever, when he said :

" Go ye, therefore, and teach all nations, baptizing them in the name of the Father, and of the
Son, and of the Holy Ghost." This practically
says: " Go, make converts, and seal them to the
new life equally in the name of the Father, the
Son, and the Holy Spirit." Now whether any
human expression of that doctrine is true or not,
evidently by his use of these names our Lord
understood them in some real sense to refer to
the One God.

Exactly analogous is the benediction in 2 Cor.
xiii: 14—" The grace of the Lord Jesus Christ and
the love of God, and the communion of the Holy
Spirit, be with you all." Here St. Paul associ-
ates the three names in the same formula, and
apparently teaches that what is back of one name
is as real, as distinct, as enduring and as much to
be reverenced, as what is symbolized by either of
the others.

Some imagine that they detect a hint of this
truth, although only a hint, in the fact that in
the Old Testament one name of the Deity is
a plural form. "And God (Elohim) said, Let
us make man in our image." But this is a fan-
ciful interpretation. What was formerly consid-
eded conclusive evidence concerning this subject
was 1 John v: 7—" There are three that bear
record in heaven, the Father, the Word, and the
Holy Ghost: and these three are one." But that
is spurious. It was probably interpolated into the
sacred text by some scribe more anxious to sup-

port a theory than to transmit truth. The words are omitted in the Revised New Testament.

As these three names are joined in the final commission of Christ to his disciples, so all through the New Testament with each name is associated the same powers and attributes. In the first chapter of John's Gospel we read: "In the beginning was the Word, and the Word was with God, and the Word was God." The "Logos —the Word—an expression used by the Greek philosophers to represent the utterance or manifestation of the Divine Reason,—is used by the Apostle John to represent God as manifested in a distinct personality, intelligible to man." The Word of God, in John's Gospel means the Son of God. God is thus said to have created the world by the Son. The Son is declared to reveal the Father. The Son himself said: "He that hath seen me hath seen the Father." The judgment is to be in the Son' shands. In the same way, the same epithets and attributes are ascribed to the Holy Spirit. The most superficial reader of the New Testament must have noticed that at times Christ is spoken of as if he had the power and authority of God; and that similar language is applied to the Holy Spirit. Sometimes the three names are used interchangeably. These names are associated in the most sacred relations. In the last utterance of our Lord, already given, they are classed together in such a way as implies

equality:—Go preach and baptize in the three-
fold name. If that were the only reference in
the Scripture to this subject I should hail it with
the utmost delight, for to me instead of being a
simple mystery, of no use except to cause good
people to stumble, it is a light which makes clear
and reasonable the otherwise incomprehensible
mystery of God. The problem, then, is simply
this: Are these three names for One Being, or
do they denote three distinct persons?

Having thus before us in outline the Scripture teaching on this subject, observe a few fundamental facts, as applicable to it.

There is only one God. The Trinitarian believes in one God, and only one, as truly as the
Unitarian. It is not a question between one and
three: but between one who is a simple person,
and one who is more complex, as a man is more
complex than a plant. The Unitarian says, God
is a person like ourselves, only infinite of course:
the Trinitarian says, He is one as we are only not
a simple but a complex personality. Life ever
moves toward complexity, and the highest is most
complex. Keep this clear,—Unitarian and Trinitarian *alike* and *equally* believe in one God, and
only one.

These names in the New Testament are all
applied to God, therefore they either represent
three distinct beings, or they are three distinct
names for one Being. But names of inferior
beings would not be applied to the Supreme:

the presumption is, therefore, that the names are
different appellations for one Deity. Jesus himself, quoting "the first and great commandment," declared, "The Lord our God is one."

The chief difficulty on this subject has arisen
from the unfortunate phrase, which is not Biblical, "Three persons in one God." But the word
"person" is an expedient of philosophy. Philosophy tried to explain the inexplicable, and
took that word "person" for a crutch, and then
limped all the way through its explanation. This
word is misleading, whether used by a Unitarian
or Trinitarian. The Trinitarian says, "three
persons in one person;" and the Unitarian re-
plies, "That is absurd! how can one be three at
the same time it is only one? You believe that
God the Father is one person, and the Son is
another person, and the Holy Spirit is another
person: that each has a separate work, and yet
that there is but one Person,—the idea is incon-
ceivable!" Yes, and that word "person" has not
the slightest Scriptural foundation ; but it is used
by theologians with careful qualifications. My
studies in theology were under the most eminent theologian who has ever lived in America,
and he was always careful to say that by "person"
we really mean *distinction*—three distinctions in
one person, rather than three persons in one
person. Professor Park's words* are, "There is a

---

* From reports of his Lectures as delivered to his students,
1869.

threefold *distinction* in the Godhead . . . . and
these three are one God." Again he says : " The
word ' person ' when applied to the *distinctions* of
the Trinity has always been understood as
used in a peculiar sense." These distinctions
are not merely official, but eternal.

Moreover, the word " person " as used by the
Unitarian in regard to God involves as great a
mystery and as great contradiction as the word
" Trinity " when used by ourselves. All say that
God is a person ; yet God can be only figuratively
said to be a person.

"A person or agent, as we conceive the
term, . . . wills, putting forth successively new
determinations of will, without which new determinations personality is null, and no agency at
all. But God never does that. He does not de-
cide to do one thing to-day, and change His mind
and do another thing to-morrow. He has the
same purpose from eternity. So a person thinks,
and has a succession of thoughts, but that cannot be said of God, for He who knows all things
cannot think in the same forms that we do.
We remember ; to Him all things are present—
nothing has to be recalled. A person has
momentary feelings ; God is the ' same yesterday, to-day and forever.' Literally, God is not a
person, for the very word is finite in all its meas-
ures and implications, because it is derived from
ourselves. Figuratively He is a person, and
beyond this, nothing can be said which is more

definite, save that He is in some sense uncon-
ceived, a real agent who holds Himself related
personally to us, meeting us on terms of mutual-
ity, such that we can have the sense of society
with Him, and the confidence of His society
with us, *as if* He were in truth a literal person
like ourselves." *

It is said, " I believe in God who is simply one
person ; " but that statement means " I believe
in God who is infinite and finite at the same time,
who is so great that he cannot be known and yet
who is intimately known ; " and that is a contradiction as hard to get over as to say, " I believe
in God who is one and yet three." Our Unitarian friends are in the same predicament as ourselves. The difficulty for both of us comes from
trying to make simple that which in the nature of
things is complex and inexplicable. Now and
then some one asks for more definiteness in theological thought, and declares that the religious
outlook is discouraging because men insist on
saying " We don't know " before eternal mysteries. The time never was in which men had any
right to be more positive than now : and they
never were more positive about the things concerning which knowledge is possible.

Certain doctrines are indefinite as to details,
while as facts they gain emphasis from their
indefiniteness. They are too great for expres-

---

*Bushnell : *Building Eras,* p 114-115.

sion. Finite words cannot utter infinite realities.
In the Scriptures God is not defined ; the doctrine of the Atonement is not laid out as a town-
ship on a map ; the doctrine of Retribution is
left in a cloud—but a cloud full of thunderings,
and lightnings, and warnings ; and the doctrine
of Father, Son and Holy Ghost, is never care-
fully defined, and yet it is clothed with an im-
pressiveness and majesty that exact definition
never possesses.

Having thus stated the Biblical foundation
for what men have called the Doctrine of the
Trinity, and having shown that it does not
mean three distinct persons, but three distinctions in one person ; that the idea of personality
when applied to Deity involves as great difficul-
ties as the idea of trinity ; and that any revelation of what is infinite must, because of poverty
of language, be cloudy and indefinite, we now
turn to the testimony to this doctrine from reason
and experience. Some subjects are understood,
if at all, only by experience. I have learned by
my own experience what neither the Bible, other
books, nor any theological teacher ever taught
me : that in order to belief in the unity of God
the doctrine of the Trinity as stated by our Lord
is essential. Many who think they reject the
Trinity actually recognize it : many are spiritually Trinitarians and intellectually Unitarians.
Charles Kingsley was one of the most honest
men who ever lived. In a letter to a friend,

Thomas Cooper, he said : * "But my *heart*
demands the Trinity as much as my *reason.* I
want to be sure that God cares for me. . . . I
want to love and honor the absolute, abysmal
God Himself, and none other will satisfy me. . . .
and I say boldly, if the doctrine be not in the
Bible it ought to be, for the whole spiritual
nature of man cries out for it." I echo his words
—"If it be not in the Bible it ought to be."
The process by which I have reached this conclusion is as follows :

Man is surrounded by Nature, with her seas
and storms, her firmament thick-set with stars,
her earthquakes and eclipses ; day is followed by
night, and darkness is ever the hiding-place of
power. Man is a cause : he can exert energy ;
and when he sees storms careering through the
heavens and lashing the oceans into mountains, he believes the hidden energy to be in the
hands of an unseen Person. That Person man
will strive to know and to propitiate. As the
field of his observation widens there will be
larger and more awful conception of the majesty
of the unseen Being. Science whispers her
secrets—says that the stars are worlds, and that
there are galaxies of systems sailing in space ;
shows that light traverses distances impossible of
measurement ; hints at the forces which bind the
worlds together,—and thus this thought of the

---

* *Biography of Kingsley* p. 198.

unseen Person grows until it overwhelms and
prostrates. With our knowledge of the universe,
—with thought running back to the time when
the constellations were luminous and burning
mist diffused through space ; with immensity mul-
tiplied by eternity—the idea of God is simply
incomprehensible. Expedients of expression are
resorted to by baffled intellects. Some say,
Everything is God—that is Pantheism ; or, There
is nothing but eternally changing matter—that is
Materialism : but still hearts love and aspire, and
still something within keeps frantically and con-
fidently reaching for some One outside to answer
inappeasable longings. With each year of progress the difficulty increases. Science is explor-
ing the universe. The telescope opens vistas of
distance before which we are breathless ; apparently dead metals are made to palpitate with
music ; the spectroscope has analyzed the light
of stars and told what substances are burning in
their far-away fires ; it even is said that there are
stars so remote that their light traveling since
the Creation has not reached here yet, and will
not arrive until we have joined the countless
dead. God, the Creator, the King, the Father,
he who holds all things in his hands—who by
searching can find him out ? Can you imagine a
Being, who never began, and who will never
cease to be ? What do you mean by infinity ?
" Infinity " and " eternity " are words used to
veil ignorance. No man ever fully compre-

hended them. I speak the word "God."
That represents to me the First Cause of the
universe; but I do not even know what a
"cause" is, and I do not know what the "universe" is. We cannot wonder that angels veil
their faces, crying before him, "Holy, Holy, Holy
Lord God Almighty." The awful, unknown and
illimitable must always be the *first* idea of God.
Thus far there is no difference of opinion. Infinite power, infinite wisdom, infinite holiness—
that is God: and now having reached this conclusion, do we know any more about him than
before? Thus far our Unitarian friends go with
us: and here we part company.

But if we have only this knowledge we might
as well know nothing of God. He is beyond us.
Prayer is impossible, and religion a dream. We
need to know not only that God is, but that he can
be approached. If that knowledge is ever possessed it will be by his adapting himself to us.
The search for God is as old as human thought.
Much of what is miscalled "infidelity" is only
an eager cry for surer knowledge. What is God
like? Has he any thought or care for men?
Sorrow crushes; no man can help; slowly and
remorselessly fate closes around us, and the end
is not far off :—are we only grist in an infinite
mill whose stones grind on forever? Something
within tells us that the Power outside cares for us,
and that we can cry to Him and He will hear us,
—that is, we personify that Power. This is not

my experience alone: it has been universal in
all ages. Men have not only believed in God,
but they have believed in a God who has revealed
himself. Something of this idea is in almost all
religions,—indeed, must be, for if we have simply
an unrevealed and infinite God religion is impossible: there may be prostration, but there can
be no communion. I believe in God; but if I
must stop there then I might as well believe in
nothing: it were better to be blind than to see
only a force and no love. Hearts cry not only for
God, but for One who adapts himself to human
weakness. If God speaks to a man it must be in
language which he can understand: if he shows
himself it must be in a form to be recognized: if
he speaks to me it must be in the English language. There is in all men a demand which can
be satisfied only by a God revealed.

What is the doctrine of the New Testament on
this point? God answers this longing of the
soul of man in the only way it can be answered,
by *revelation in humanity*. The Logos-doctrine
is that it is eternally the nature of God to manifest himself. Whenever in the New Testament
you have a reference to God considered alone,
the name used is Father. Whenever that Father
is represented as coming into relations with men
the name is Son, or Logos (Word), that is, the
Father communicating or revealing himself—as
the uttered word reveals the thought. And so
we believe that there is forever in God that which

is manifest in what Jesus Christ was and did.
I want to know about God: his power I see in
Nature ; his feelings toward men, in Jesus Christ.
Does God care for the poor? Nature seems
sometimes to say, "Only to crush them ;" Jesus
preached the glad tidings to them and fed them.
Does God care for the sick? Jesus went about
healing diseases. Does God have sympathy for
those who have broken his laws? Jesus prayed
for those who crucified him. Does God care
for the sorrowful? At the grave of Lazarus Jesus
wept. Does God regard the masses who struggle
in sorrow and pain? Jesus said, "Come unto
me all ye that labor and are heavy-laden, and
I will give you rest." Does God really love
men? The life of Christ, the teachings of Christ,
the death of Christ, all answer,—"He seeks to
save that which was lost." Take this one phrase,
"God manifest in the flesh," and you have the
clearest teaching concerning the earthly life of
our Master.

But that revelation was necessarily limited ;
limited to a human form, by human language, to
a short period of time. What relation to God
had the millions who lived before Jesus Christ ?
What relation have we to him ? These are questions of fundamental importance. What says
Christ? " He that hath seen me hath seen the
Father." The Father does not change: God is,
always has been, always will be, in all his relations to men, just what Christ was, who died to

save men ; and so the cross declares that it has
always been God's nature to seek and save the
lost, and that it always will be.

If we continue to use the unfortunate word
" person," we may say that we have now before our
thought the first and second persons of the Trinity.

There is yet one more question. We need to
know not only that God is, and that it is his
eternal nature to reveal himself, but whether
there is in this Nineteenth Century of the Christian era, any possibility of getting near to him :
in other words, Has he anything to do with us
now, or, when Christ died and the Bible was
written, did God retire into infinite and thence-
forward unbroken solitude? It is well to know
that Christ died eighteen hundred years ago to
save sinners, but men need a Saviour now as much
as they needed him then. It is well to point to
the Bible as an infallible guide, but new questions are arising each day which every child knows
the New Testament does not pretend to answer.
The Bible says, " Love thy neighbor," but does
not tell what love will lead a man to do. We
need not only to know that God exists, and that
he revealed himself eighteen hundred years ago,
but whether he cares for us now, and whether
there is any new light to break on the darkness of
earth. A German theologian has said : " All
men have desired a human God, that is, a manifest
God ; " and I add, All men desire a present God.

What says the Bible to this? " In the begin-

ning the spirit of God moved upon the face of
the waters;"—that is, when chaos reigned, and
man was not, God was then moving on nature.
"And the Spirit and the Bride say, Come;—"
that is, when the New Jerusalem shall descend
out of heaven from God the last sound that will
echo over the old creation will be God crying to men, "Come." God, a Spirit, in personal
contact and communication with human spirits,
interpreting old truth, revealing new truth, in the
intimate daily leadership of spirits—that is the
teaching of the Bible : for that, weary, strug-
gling, defeated, but never entirely cast down, all
men continually do cry. And so we have the
doctrine of the Spirit—God always in vital touch
with the spirit of man, convicting of sin, inspiring aspiration, revealing new truth.

God pervading the universe, and so transcending thought; God not forever unknowable, but
in all things which concern men just what Jesus
Christ was; God near to each one of us, even in
our hearts, and so immanent in every soul. The
three-fold name corresponds to a three-fold demand in the human nature. We have thus before
us the doctrine as taught by our Lord :

(1) The great, abysmal, infinite One—whom
he named Father.

(2) The same One in self-revelation, from
eternity to eternity manifested through a distinct
human personality—called the Son.

(3) The same One always moving on human spirits and drawing them toward himself—called *the* Spirit.

And, now, I am ready to tell why I believe in the doctrine of the Trinity: why I can say with the poet preacher of England: "If it is not in the Bible it ought to be." It is because the universe is to me without any meaning if this doctrine be not true. Jesus Christ revealed facts; and only facts are essential to faith. Each of us is not only justified but required as best he can to harmonize facts. Others have no more right to tell us how we must interpret facts than we have to tell them. The simplest statement of any doctrine is usually the best. The words of Christ in his last commission are clear; he told his disciples to baptize in the three-fold name, because that condensed the Gospel. I believe in that doctrine, because the Bible teaches it, and because the human soul demands it.

I believe in God, the Father Almighty.

I believe that it is an eternal distinction of the nature of God to manifest himself, and that in Jesus the Christ he has so manifested himself to man.

I believe that God did not wait until Christ came, to have intercourse with men, but that he had been moving on human hearts in all ages and lands, in China and India and on the Islands, as well as in Judæa, and that since Christ has gone He has been continuing and will continue forever this work of salvation.

Thus the world's longing for God, for a human God, for a present and never-leaving God, is met and answered in the three-fold name. Is this simply a confusion of words? Is this simply a theological conundrum? None who have seriously considered the problems of human thought and life can think that it is. This doctrine of God is the only answer that I know to our profound and otherwise unsatisfied longings. I am afraid that many wander in darkness because, while they vaguely believe in God, they have not yet seen him as revealed, and as ever-present. " It is one of the merits of this doctrine," says Bushnell, " that it does not fool us in the confidence that we can perfectly know and comprehend God by our first thought." To me—as our Lord stated it, not as men have philosophized about it when trying to stretch it on the iron frame of their theological systems—the doctrine condenses the whole Gospel ; it is the door through which the new day streams upon the darkness of humanity ; it is the truth to live by, the truth to die by, the truth that makes immortality a necessity.

And now what can we do, who rejoice in this superlative reality, except key our hearts and our lives to the sublimest of doxologies, which has been chanted with fervor and harmony continuously growing through all the Christian centuries, and which will be chanted until the earthly

music shall break into the " Holy, Holy, Holy,"
of the Heavenly temple—"Glory be to the
Father, and to the Son, and to the Holy Ghost;
as it was in the beginning, is now, and ever shall
be, world without end. *Amen.*"

III.

WHAT IS LEFT OF THE BIBLE?

" I find more and more that the Bible is made very little use of for its true purpose, but all the more for purposes which are quite foreign to it."—ROTHE.

" God be praised that this our book has not the clearness of a symbol or creed: that we are not forced to comprehend it aright, and that we may give many meanings to his word . . . . It is not in a chain of dry sentences that God reveals to us his will and the principles of his government: it is essentially by facts."—ALEXANDER VINET.

> " Out from the heart of nature rolled
> The burdens of the Bible old:
> The litanies of nations came,
> Like the volcano's tongue of flame,
> Up from the burning core below,—
> The canticles of love and woe."
>
> EMERSON, *The Problem.*

" Before all things the Bible is the book which has reached the highest conception of God yet attained by the human consciousness."—HENRY WARD BEECHER.

" The redemptive purpose of God was not ushered into the world a full-grown fact; it evolved itself by a regular process of growth, and the process was marked by three salient features: slow movement, partial action, and advance to the perfect from the more or less imperfect, not only in knowledge but also in morality."—PROFESSOR ALEXANDER B. BRUCE.

> " Word of Mercy, giving
> Succor to the living;
> Word of life, supplying
> Comfort to the dying."
>
> H. W. BAKER.

" The words that I have spoken unto you are spirit, and are life."—JOHN vi. 63.

III.

What is Left of the Bible?

THE air is full of discussions concerning the Bible. Those who would discredit it have at least one great problem before them, and that is, How a common book, written by ordinary men, can have commanded, for so long a time and under such varying circumstances, the attention and study of such different classes of people. Quite as many eminent and able scholars are devoted to its elucidation as to the sciences of geology or astronomy. Those who do not recognize in it any binding authority, still, by some subtle attraction, spend years over its pages. Some study it that they may overthrow it ; others that they may discover its place in the world's literature ; others that they may use it in support of systems of theology ; others for the splendor of its literary style ; and vastly more because of the light which it sheds on this life and the hints it gives concerning the life to come. Even great scientists, like Huxley and Tyndal, turn from their

laborious investigations and seek recreation in
what may be called "theological and critical by-
play."

And this is no new thing. The Bible for a
thousand years has been more in the thought and
attention of the civilized world than all other
books combined. That it has drawn the attention of such acute and finished destructive criticism is a majestic tribute to its greatness. This
study of The Book seems to have culminated in
our time. Scholarship revolves around it : and
the profoundest thinkers of the Old World and
the New are attempting to solve the problems
which it suggests. Many earnest and timid souls
who do not yet realize that the foundation of our
faith is not in the dead letter but in the Living
Spirit, not in the miracles of two thousand years
ago but in the continuous miracle of all the two
thousand years, feel as if the whole structure of
Christianity were falling.

We have heard much during the past year or
two concerning critical questions—Who wrote
the books of the Bible? and, When were they
written? I desire to direct your attention to certain facts which cannot be changed by any theories concerning the authorship of the books, or
the structure of either of the Testaments.

In the first place, observe that not one of the
books whose authorship is called in question as-
serts that it was written by the person whose
name it bears. Tradition says that the first five

books of the Old Testament were written by
Moses. If new light shall prove that they were
written by some one else, the world loses nothing. That they have been believed to be of
Mosaic authorship no more affects their teachings than the quality of the metal in its case affects the accuracy of a watch. The time was
when the Ptolemaic Astronomy was believed to
be infallible; and when a more reasonable suggestion was ventured millions of pious souls were
sorely troubled. But the change came, and instead of the evil effect which had been predicted
the universe expanded, and the idea of God be-
came more glorious and inspiring. I make no
defence and no condemnation of Higher Criticism. A great change in the interpretation of important passages of the Bible may be forced. If
it ought to be, the quicker it comes the better; if
it ought not to be, clearer light will end in more
firmly establishing the old positions. Never were
so many consecrated Christian scholars studying
Holy Scripture. If we cannot trust that they
and those like them will by and by reach truth,
we can believe nothing. Those not experts in
the languages in which the Bible was written
should remember that their opinions on these
questions are not worth the words taken to utter
them. It is ludicrous to suppose that good men
are necessarily wise critics, or that questions
which only careful scholarship can answer can be
settled by a show of hands. Great scholars are

studying the Scriptural problems, and great scholars are on both sides. Whether Moses or some one else wrote the first five books of the Old Testament ; whether there were one or two Isaiahs ; whether Paul, or Apollos, or some un-known writer is the author of the Epistle to the Hebrews, is of little importance so long as the truth which the books contain, and which is cer-tified by its correspondence with the needs of humanity, remains untouched.

" We believe that a church is a society of men possessing the life of the Eternal Son of God, and having direct access through him in the power of the Spirit to the Father ; of men know-ing for themselves, at first hand, the reality and glory of the Christian redemption ; of men to whom the truth of the Christian Gospel is au-thenticated by a most certain experience, the ex-perience not of the individual life merely, but of a society.   Is this consistent with the agitation, the heat, the panic, created by the assaults of critics on the histôric records of the Jewish and the Christian revelations ?   We of all men should keep calm.   These controversies leave untouched the strong guarantees of our faith.   For every church is a society of independent witnesses to the grace and power of Christ.   For us the im-mediate manifestations of the eternal life that dwells in Christ are found not merely in the words and deeds and sufferings recorded in the Four Gospels, but in the company of the faith-

ful. We know that Christ is alive from the dead,
for He lives in them."*

Let us now attempt to answer the question,
What would be left of the Bible even if all the
claims of the critics were valid? Is there any
vital truth in the Holy Scriptures which is independent of theories concerning the authorship of
its books, or their arrangement in the canon?

The Bible is the Book of *God.* At its beginning we meet that word, and never does it disappear. The creation is ascribed to God—"The
spirit of God moved upon the face of the
waters;" when man appears he is in the image
of God ; when he has fallen it is not out of the
care and watch of the Almighty, for from beginning to end echoes the Psalmist's exclamation—
"If I ascend up into heaven thou art there: if
I make my bed in Sheol, behold, thou art there.
If I take the wings of the morning, and dwell in
the uttermost parts of the sea, even there shall
thy hand lead me, and thy right hand shall hold
me." The world is represented as resting in
God's providence. Enoch walked with God ;
Abraham was the friend of God. God planned
for the education of the race in spiritual things.
To realize that purpose one nation was chosen.
That nation was selected not to be the only one
saved out of the world, but for the purpose of

---

* R. W. Dale, *Address at The International Congregational
Council, London,* 1891.

saving the world. The Jewish people were
separated from others that they might be a light
in the darkness. The light is no more favored
than those who walk in its rays. That the
millions in heathenism, as well as the few Hebrews, were not outside the providence of God is
evident because he prepared a peculiar people to
make visible his nature and presence in the
world. The story from Genesis to Revelation
thrills with one thought;—men and nations, in
all conditions and times, rest on the heart and in
the hands of infinite and never-failing Father-
hood. At the beginning God creates: at the end
the Spirit invites. At the beginning are the
clouds and storms of a weltering chaos, but above
them God; at the end the glory and splendor of
the New Jerusalem descending out of heaven
from God; and in what may be called the center
of the space between is uplifted the cross on
which the blood was shed, revealing that even
God enters into suffering and sacrifice in order
that his purposes of blessing may be accomplished. To take God out of the Bible would be
like taking the sun out of the firmament. Before
the storms and earthquakes, before the bright-
ness of the sun and the beauty of the stars, the
heathen have bowed; but the Bible, retaining
all the glory and majesty of the idea of God
which comes from the creation, adds to it the
assurance that from the beginning until now
humanity has been held in the leashes of an

infinitely loving heart. With that thought our
Bible begins, and with that it ends. No one can
escape from the love of God. It is all-embracing,
like the atmosphere; all-pervading, like the light,
without beginning of years or end of days—
for the cultivated and ignorant alike. More
than that, even the creation itself "groaneth and
travaileth in pain, waiting for the adoption of
the sons of God." The greatness of this truth
passes comprehension. It is high; who can
attain unto it? All men, all nations, and even
the physical universe, are encompassed by infinite and never-ending Love,—that is the note
to which is keyed all the music of our Bible.

The Bible is the Book of *Forgiveness*. Love
would be mockery without forgiveness. He who
carries in his heart a guilty secret imagines that
the universe is armed for his destruction: even
the physical forces seem to be his enemies; he
hates the light. "Evil loves darkness." There
is a profound philosophy in that phrase. How it
came to be true we need not inquire. Whether
there is a personal spirit of evil, or whether sin
is natural weakness, are questions of little importance. In all ages, and among all peoples,
the consciousness of guilt has been terribly real.
The system of sacrifices had its rise—no one
knows when—in the consciousness of man that
he had wronged the Supreme Power. Whether
that power is God, or the Devil, the longing for
forgiveness is manifested everywhere. You might

as well try to take out of the world the Alps or
the Andes as this consciousness of guilt. This
fact the Bible recognizes, and, almost alone
among books, addresses itself to answer the
inappeasable and universal hunger for pardon.
The Hebrews at first were allowed to offer sacrifices, probably because they were essential to the
popular idea of religion ; but, gradually, as virtue
was attached to them rather than to God, the
warning voices of the prophets were raised, and
we hear the cry: "What doth the Lord re-
quire of thee, but to do justly, to love mercy
and to walk humbly before thy God?" That is,
the sacrifice which is pleasing to God is not
fruit, wealth, the blood of animals—these are only
symbols—but righteous acts and holy character.

The revelation of God as the forgiving Father
culminates in our Lord Jesus Christ and in his
supreme message: "If we confess our sins He
is faithful and just to forgive us our sins, and to
cleanse us from all unrighteousness." Men had
asked, "What shall we do that we may appease
the powers above?" To stifle remorse they had
gone on pilgrimages, given their choicest treas-
ures, even slain their children with their own
hands, yet had found no peace. But from the
time when the Hebrews were chosen until the
cross was lifted one truth shines with ever clearer
radiance:—God is not a Being who conditions
forgiveness on suffering; the sacrifice that is
acceptable to him is repentance, turning away

from evil, a new spirit and a clean heart. And
so when men ask, What shall we do that we may
escape from the consequences of sin? the Divine
word is, You cannot escape. If you trust to
yourself you must go on suffering forever, but if
you turn away from your sin unto the Living God
" He is faithful and just to forgive," and will in
due time remove even the evil desire. You
might as well try to get the color out of a sun-
beam as the thought of forgiveness out of either
the Old Testament or the New. Whatever may
be the result of critical inquiry concerning the
structure of the Book, the great voice with which
the Bible closes sounds on forever :—" Whosoever will, let him come, and take of the water of
life freely."

The Bible is the Book of *Righteousness.* Its
offer of forgiveness never for a moment obscures
its emphasis upon right living. All its teaching bends toward the perfection of character.
The Hebrew bowed before Jehovah and cried,
" Holy, Holy, Holy"; Isaiah in his vision saw
the Lord, and heard the seraphim crying one to
another, " Holy, Holy, Holy is the Lord of
Hosts; the whole earth is full of his glory "; and
the Master, when he was condensing the whole
duty of man into one never-to-be-forgotten sen-
tence, said, " Be ye holy, as the Lord your God is
holy." Men are always like the Deity they worship. This idea of a holy God separated the
Hebrews from all other nations. The Greeks, not-

withstanding the splendor of their literary culture
and artistic development, believed in gods who
were immoral, cruel, heartless, vindictive, despica-
ble. The depravity of heathenism has been to
think of God, or the gods, as possessed of few of
the virtues and all of the vices of humanity. High
above such degrading conceptions of Deity was
lifted the ideal of the Hebrews. It has no finer
illustration than in the Book of Job, whose music
is keyed to the righteousness of God. Job's
friends declare that he has sinned, and, therefore,
is being punished ; but Job in conscious integrity
refuses to believe in a Being who would punish
one not guilty. He prays that he may die rather
than give up his faith in a God of absolute
justice. It has been said that the Book of Job
shows us, not a man before the judgment-seat of
God, but the Almighty before the judgment-seat
of man. Job's cry has been echoed in Whittier's
familiar lines :

> " The wrong that pains my soul below
>   I dare not throne above."

That glorious drama of God and man is no exception in the Old Testament. Toward a life of
righteousness all men are pointed. When they
choose evil they are denounced. The mask is
torn from the hypocrite Jacob, the beautiful gar-
ments from the lecherous David ; even Solomon
is shown to have lacked an appreciation of true
wisdom ; and when the Perfect One appeared it

was in the realization of perfect righteousness—
" Who did no evil, in whose mouth was no guile ;
who went about doing good, and the law of
whose life was love." In the Bible bad men are
frequently described, but never with approval ;
infamous and cruel acts are recorded, but never
commended ; and, if they are endured for a time,
it is only that the higher end of justice and truth
may be realized. That cannot be wrong, that
cannot be from beneath, that cannot be simply
human, which begins and ends in the holiness of
God and the righteousness of man.

The Bible is the Book of *Life*. This fact is
the cause of much misconception. We are so ac-
customed to think of books as made that it is
hard to understand that they can grow. But the
Bible is a growth rather than a mechanism. It
contains no propositions, and nothing like a theological system. There is not a hint that any one
of its writers, not even the Apostle Paul, ever
heard of such a thing as a logical process. A
part of the Old Testament is a history of God's
dealings with the race ; other parts record the
utterances of prophets who spoke the words of
God to the men of their time ; others still are
hymns of praise and prayer—the rarest lyric
music that ever broke upon human ears, voices of
the day and voices of the night, cries of despairing spirits, longings for communion with the unseen, hisses of execration and wails of despair,
songs of confidence and hope. The music of the

Old Testament was not made to order; it has
come from the heart of man beneath the Spirit's
touch like the water from the rock when struck
by the wand of Moses; it has sprung like the
gushing of fountains from perennial springs.
Other parts are condensations of the wisdom of
the time, like Proverbs and Ecclesiastes; at least
one book is a great dramatic poem—at once a
drama and an epic; for Job has been fittingly
called "The Epic of the Inner Life." And then
there are visions, wonderful openings from out of
the darkness of the present into the splendor of
the future.

All this is in the Old Testament. When we
come to the New each part is instinct with life.
"The life was the light of men." What human
teachers would attempt to convey by prosy
books the Divine Teacher has conveyed by a perfect Man. The New Testament is the record of
a life, and meditations on that life. Take away
the life, and nothing remains. The Gospels
describe the life in the flesh; the Acts the beginning of that life after the body had been laid
aside; the Epistles are meditations on the same
life as it grows in the midst of an effete and per-
nicious civilization.

To attempt to read the Bible as you would read
a treatise on Ethics or Logic is to misunderstand
its meaning. The lesson of Enoch is that those
living to-day may walk with God; and the lesson
of the story of Joseph is of the providence of

God in human affairs. A literary critic might
read the Psalms and find them only common
poetry, but the man with a great joy or a crush-
ing sorrow finds them voices from beyond the
stars. One fact separates the Bible from all
other books—it is occupied from beginning to
end with life; it is written in terms of life ; its
messages are conveyed through human beings
and never in logical propositions. It is, therefore, adapted to all classes; the wisest can never
exhaust life, and the poorest are always teach-
able in its presence.

Not only is the Bible vital ; it is also vitalizing.
It inspires life. Those who turn to it to find
specific rules for their duty will be disappointed,
but those who come in contact with the spirit
which dwells within it, as a spirit in a body, are
unconsciously transformed. He who said that
the Bible was to him not so much a book of rules
as of germs, uttered a great truth ; for its words
fall into hearts and grow, and are valuable not so
much for what they teach as for what they sug-
gest. It exhausts no subject ; its principles can
be tested only by living them. No one knows
what love is without loving, what goodness is
without being good, what the peace of God is
without resting in that peace. The messages of
Scripture are conveyed through fallible men, in
ordinary human language, and in that fact is their
power. It would be a poorer book if it were
more finished. If its writers had attempted to

make it perfect it would have been meaningless
to the common people. It is like nature, which
philosophers study forever without exhausting,
and which peasants enjoy forever without being
weary. It is a book of life, and it inspires life.
Those who are true to it are always loyal to conscience, righteousness, love and hope ; and everything which binds a man with golden chains to
conscience, righteousness, love and hope, binds
him to the throne of God.

The Bible is the Book of *Hope and Promise*. In
the beginning a small ray of light appears, which
expands as it extends, until the prophecy—" the
seed of the woman shall bruise the serpent's
head "—finds its fulfilment in Him who broke
the barriers of class, sect, nationality, and gave
new emphasis to the truth that all men are the
children of God, and brothers one of another.
Throughout the Old Testament this note of hope
is never lost. Prophets at times seem to utter
voices of despair, but even Jeremiah beholds the
approaching dawn, while the prophecies of Isaiah
ring like anthems. They begin with denuncia-
tions of wickedness, and end with visions of the
time when nations shall learn war no more, when
every man shall sit under his own vine and fig-
tree, and there shall be none to molest and make
afraid. The Master's message culminates in
hope. The world lies in guilt and sin, but he
will break the power of sin. Death casts its pall
over a despairing race, but he declares that the

universe is the Father's House, and that dying is
only going from one room to another in the infinite palace. His words are taken up by the
Apostle, and we hear them echoing among the
heights of his transcendent chapter—" Death is
swallowed up in victory " ; while the canon closes
with the most splendid and inspiring glimpses of
a time in which there will be no more sorrow and
no more pain ; where the Hand that lights the
stars and leads the constellations, wipes away all
tears ; and where all beings live and see in love,
as now we live and see in light. I say without
fear of contradiction, there never was another
book so full of hope as the Bible. Even Job
catches glimpses of the Daysman—the Mediator
—through whom he can approach the Almighty ;
and his question, " If a man die shall he live
again ? " is more than inquiry—it is the faint
but eager assertion of a truth dimly but surely
seen.

Read the Bible : become saturated with it :
take its words concerning God, humanity, the
forgiveness of sins, and what comes after death,
and then be a pessimist if you can ! The Scriptures represent all things as moving toward
higher and finer conditions. The processes may
be slow, but the triumph is sure. The Bible is
the world's Book of Hope.

The Bible has *Jesus Christ* in it. We believe
in the Bible because Jesus Christ is in its pages.
Adopt any theory of the structure of the Old and

New Testaments that you please; imagine that
Moses, or some one else, wrote the Pentateuch;
that there was one Isaiah, or a dozen; that John
wrote the Gospel which bears his name, or that
it was written by an unknown Alexandrian; let
all these theories have whatever weight may
belong to them, let the analysis go on—what difference does it make with the contents of the Scripture? The Old Testament points toward the
Coming One, and the New Testament is a record
of the One who has come. Suppose the story of
Jesus is a fiction—what then? Then a fiction is
the mightiest power that ever broke upon the life
of men. If Jesus Christ had his origin in the
minds of some obscure Judæan fishermen, then
those Judæan fishermen were the greatest men
that ever lived, since they have created for the
world its Saviour—nay, they have created its
God. Whether Jesus is a fiction or a reality he
is the world's Saviour, and is gradually bringing
in a higher and better order. His power on life
is independent of the question whether he is fact
or fiction.

But the reality of his life no intelligent man
questions. Doubtless there are many untrue traditions about him, and possibly some have found
their way into the Gospels—what does it matter?
One colossal character rises in the midst of the
Bible and casts his light over all the centuries.
There has been one Man concerning whom criticism is dumb; One has lived a faultless life;

One has proved that righteousness is possible ;
One has taught that love is the highest law ; One
has brought answers to the questions of humanity which perfectly satisfy all our most eager and
persistent inquiries ; One has revealed God as
Father ; One has said that the law of righteousness is the law of love ; One has declared in a
voice which the world has heard and heeded that
all sins are already pardoned, and that whosoever
will may enter into the possession of forgiveness ;
One with unsandaled feet has walked through
the valley of the shadow of death and come up
into immortal life radiant and triumphant.
Doubt may stand before him and question.
Timidity may say, " Oh, if it were only true ! "
Unbelief may say, " We will have none of him."
But still he remains, filling both Testaments,
gradually becoming " the master-light " of all our
seeing and the master-thought of all our thinking, lifting human hearts out of gloom and despair, and opening before all the gates of hope, of
peace, of " far-off infinite bliss." He is making
all things new, bringing in a better social order,
creating new States, giving new meaning to the
long-forgotten doctrine of Brotherhood, and lead-
ing all men and nations to things above.

We are in the midst of times which try men's
souls. Investigation is pushing itself into all
realms ; science is knocking at all doors ; there is
no more a Holy of Holies in all the temple of
nature ; things which have been held sacred are

fading as the leaves ; old traditions are disappear-
ing ; opinions and beliefs which have been supposed to be important, no longer have authority.
No wonder that many are asking, Where will all
this end? There are even those foolish enough
to say, "Let us cling to what we have had: let
us shut out the light because if we let in the day
something we have held sacred will disappear."

I bring to you a different message. Remember
that nothing true can ever be weakened by truth
and light. Let in the light! let it penetrate
every dark place! Truth has nothing to fear
from investigation: only error is cowardly.
Darkness never loves light. And while the process of criticism goes on, remember that whatever
may be found true concerning the structure of
the Bible, and the authorship of its books, it will
remain forever—the Book that is filled with God,
as the bush before Moses was filled with flame ;
the Book of Forgiveness, making plain the way
out of sin into life and peace ; the Book of Righteousness, showing that all things in the heavens
are in the hands of perfect Goodness, and that all
things on the earth move toward "a far off Divine event"; the Book of Life, written in terms
of life so clearly that little children can understand it and sages never transcend it, whose
words linger in the memory like seeds in the soil
of springtime, and grow and bear fruit in lives of
truthfulness and love ; the Book of Hope, in
whose pages humanity is represented as moving

ever toward a brighter and clearer day, in which the life of man is not described as ending in oblivion and nothingness, but growing in beauty and power, deathless as the Being from whom it came. And this Bible, which is the Book of God, the Book of Righteousness, the Book of Forgiveness, the Book of Life, the Book of Hope, is finally the Book which thrills and throbs with the radiant presence of Him who said: "I am come that ye might have life, and that ye might have it more abundantly"; who said, " Him that cometh to me I will in no wise cast out"; who said, "A new commandment I give unto you, that ye love one another as I have loved you "; who said, "In my Father's House are many mansions; if it were not so I would have told you."

That Book is the world's book: it belongs in the list of elemental facts, like the suns and the stars. We can trust it while we live, trust it when we die, and we shall find its messages all true when we have gone through death into the clearer light and the cloudless day.

IV.

THE IMMORTAL LIFE.

"Whether we be young or old,
    Our destiny, our being's heart and home,
    Is with infinity, and only there."
                                        WORDSWORTH.

"The grave is but a little hill, yet from it how small do the
great affairs of life look, how great the small!"—THOLUCK.

"Life is short, death is certain, and the world to come is everlasting."—JOHN HENRY NEWMAN.

"The disposition to disparage the 'personal life,' and let it go
as an 'individual accident of the universal' probably arises from
an unconscious confounding of it with the bodily form: on the
break-up of which the spirit was supposed to have no retaining
walls, but to escape as a vital breath and mingle with the general
air."—JAMES MARTINEAU.

"How pure at heart and sound in head,
    With what Divine affections bold,
    Should be the man whose thought would hold
    An hour's communion with the dead!"
                                        TENNYSON.

"If a man die shall he live again?"—JOB XIV. 14.

IV.

THE IMMORTAL LIFE.

IN one of the chapels of Westminster Abbey are buried the remains of a noble woman. Year after year loving hands have placed upon the cold marble on which her name is chiselled beautiful and fragrant flowers. I have never looked upon those flowers without feeling that they were both a tribute to the one who had gone and a silent, and almost resistless, argument in favor of continuance of life. While she was on the earth she often walked through the chapels of the dim old Abbey, for her home was almost beneath its roof; and yet the Abbey stands, and will for generations to come, but Lady Stanley, who won her way into human hearts by beneficent ministries, "and led all to things above," is supposed by many to lie beneath the stone that closes her tomb. Then a tree is more enduring than a heart: then those who make the life and beauty of a house are less than the house itself.

On a table in a parlor is a photograph, and

before it a vase of fresh flowers. On a wall is a
large portrait, and its frame is twined with a living vine. The flowers and vine are more than
tributes to the past, they are witnesses to a faith,
however feeble, that those who have gone still
live and do not refuse these tributes of affection.
And what a strange thing is a vase of flowers
before a photograph! The photograph may last
for a century, or a thousand years, and not one
lineament of the face disappear; and is man less
enduring than his photograph? Does he die;
and does his portrait which the sun prints on a
piece of perishable paper have almost the quality
of immortality?

Job had seen everything go. An evil disease
had taken hold of him; his friends had denounced
him as a sinner; he had nothing left but the consciousness of integrity; he had even prayed that
he might die rather than be untrue to his faith in
a just and righteous God;—when, suddenly, a
thought flashed upon him: "If a man die shall
he live again?" That is more than a question:
it is a faint but real affirmation—a voice declaring
that while there is no explanation of life as it is,
if we may believe that it continues, all its mysteries may be explained. That old poem might have
been written in this nineteenth century with
scarcely any change. It is a profound and true
study of human life, the problems of which in one
time are the problems in all time. Other things
have changed. More is known of the universe

than ever before. Progress has been multiform.
The very stars have found voices; the light has
become vocal; the secrets of the deep seas and
adamantine rocks are being read; but in the nineteenth century after Christ there is no more
known concerning life and its mysteries than was
known in the nineteenth century before Christ.
The question of Job is world-wide and world-old.
I bring to you no new light, but rather gather up
a few scattered rays and allow them to flash their
gleams on this endless study.

Jesus came to a people ready to welcome death
as the only way to escape from the miseries of
existence. He never recognized death. Before
he healed the daughter of Jairus he said, " She is
not dead, but sleepeth; " and when he came to
the grave of Lazarus he said, " Our friend Lazarus
sleepeth." His last words were, " Father, into thy
hands I commend my spirit." His answer to
Job's question is a strong and unhesitating affirm-
ative. For a moment, however, let us turn from
our Master and listen to other witnesses.

The immanence of God gives authority to the
voices that speak in the human soul. If God
pervades his universe and is in humanity, then the
voices of the human soul, when they can be heard,
have the authority of the word of God; and truth
which can be found anywhere, even if not written
in the Scriptures, is not to be ruled out as belong-
ing to an inferior realm, but it, too, comes to us
with Divine sanction. I do not feel that the

only light upon the mystery of death is found in
the teachings of our Master. He treats the subject as he treats the idea of God. The existence
of God is presumed, never proved, in the New
Testament. In the same way from beginning to
end Jesus presumes continuance of life. The
importance of that phrase, "If it were *not* so I
would have told you," cannot be exaggerated.
It recognizes other witnesses whose testimony is
entitled to reverent consideration. To their
utterances we now turn.

The hypothesis of future life makes possible a
solution of the mystery of this life. This thought
is prominent in the Book of Job. It has forced
itself into the thinking of all time. The wicked
succeed and the good fail; the bad are exalted
and the pure are cast down; the vicious have
palaces and the virtuous live in hovels; Domitian
wears the purple and Epictetus is lame and a
slave; Nero sways the sceptre and Jesus hangs
on the cross. Such sights are not universal, but
they are so common as to confuse and confound
our thinking. Why is sin allowed? Why are
millions born into conditions which necessitate
evil? One little outcast child—with neither
father nor mother, no school but the street, no
home but the brothel, no companionship but that
of thieves—is an awful mystery. Job could see
no meaning in life. He was ready to die, until
there came to him the thought of a life beyond
the grave. If there is life beyond, he seems to say,

then what now is wrong may be righted. What helped him, illuminates the same mystery for us. Evolution points to a perfected race in far-off ages, a race which may sometime be immortal, but has no answer when we ask, How about those living now, who will never know anything of that perfected race? Are they any more than the blossoms which never come to fruitage, and the weak who go down in the struggle of life? There is no possible solution of the mystery except on the supposition of continuance of being. Professor Fiske well says at this point "faith must appear and bridge the gulf."

If life endures, there is a motive for the development and exercise of the noblest of human faculties. Our best faculties become free and efficient only through discipline and struggle. The noblest who ever lived was "made perfect through suffering." But now the question rises— and it is intensely practical—if the best and noblest can be realized only through struggle, sacrifice, sorrow, and can be used but a little time, "Is the game worth the candle?" Would Raphael have painted the Sistine Madonna if he had known that as soon as its divine beauty was perfected a remorseless hand would tear it into a thousand shreds? Would Germany have finished that almost ideal temple which lifts its splendor on the banks of the Rhine had it been known that, within five years after the crosses were placed upon the spires of Cologne, the whole mass would tumble into ruins? Why should any seek improvement, with its accompaniments of anguish and agony, if death is at the end of the way? Why should any try by long and painful culture to perfect characters which, almost as soon as the process is ended, by a ruthless fate will be utterly and irremediably annihilated? But if life continues then there will be a sphere for the use of those faculties. Man's passion for advancement, culture, perfection, is explicable only on the hypothesis of unending life. While I fully believe that there is more happiness in virtue than vice, in culture than in ignorance and neglect, I cannot fail to recognize also that, theorize as we may, to most the motive for improvement will go if all hope for the future is lost.

If death ends all there is no explanation of the fact that most die at the very time when they are best fitted to live. The list of great ones who have fallen in the midst of illustrious promise is long indeed. Shelley, whose music was like that of his own skylark, finished his singing at twenty-nine: Keats "felt the daisies growing over him" at twenty-two; Frederick Robertson was only thirty-eight when his voice was hushed: Professor Clifford had ceased his investigations at thirty-four: Sister Dora was only in middle-age when death claimed her. More than one building at Harvard, Yale, Amherst and Princeton, is a memorial building. Young men pursue their studies, give brilliant promise, and just as great things are expected of them strength fails and tragedy ends the farce. A fireman climbs a lofty building to save a child. He is young, vigorous, brave, and full of hope. The child is saved, and the man lost when most proving his right to live.

What is true of youth is also true of age. Old age by no means implies decadence of power. The Apostle John did his best work after he was seventy. The controlling statesmen of Europe are nearly all sixty years of age, or over:—Caprivi is sixty-one, Crispi over seventy, Bismarck seventy-seven, Salisbury seventy, Gladstone over eighty, Castellar sixty, and Pope Leo not far from eighty-two. Of modern historians, Bancroft and Von Döllinger continued to write when they were far beyond eighty; Bryant wrote his "Flood of Years" when he was eighty-two; and Longfellow was only about ten years younger when his muse rose to its loftiest flight in the Alumni Poem at Bowdoin College entitled, *Morituri Salutamus*. England's Laureate is singing as sweetly as ever, at eighty-two, and our own Whittier, about the same age, like the Apostle on Patmos is still "in the spirit," and still sounding the praises of the "Eternal Goodness." These men will not die because they have ceased to be of service to the world, but death will find them with their spirits' eyes undimmed and the strength of their spirit unabated. If there is continuance of being, those who seem to die only

change their sphere of activity: but if there is
no continuance of being, the brightest, finest,
most splendidly equipped of our race drop into
nothingness at the moment they are best fitted to
be a blessing and a joy. To believe that, is to
deny every voice that speaks in the human soul,
to put shackles on reason, and to say that that
which seems impossible is the only reality.

If there is no life beyond death, we must
believe that humanity is created for no high purpose. There can be no noble purpose if the
grave is the end. If men are only like the drops
in the river rushing toward Niagara, disappear-
ing forever when the abyss is reached, why
do any live? Live to love and be disappointed,
to aspire and fail of achievement, to have a few
blessings and then to lose them ; live long enough
for dear ones to get entangled in our heart-
strings and then have our heart-strings torn to
shreds ; live to build houses and plan for enjoy-
ment and awake to find the dream broken? But
with Job's hypothesis the end is not yet. That
which seems without purpose or plan moves
toward a time in which there is opportunity for
great things. If we live but to die, there is
no possibility of an adequate purpose in any
human being ; but if death is a sleep by which
worn powers are refreshed that the new day may
be the better enjoyed, all things fall into harmony.

Moreover, if life survives death the voices of

the human soul prophesy what will be realized. What poet ever yet sang all the songs which he heard? What painter ever painted all the pictures he has seen in the halls of his imagination? What musician ever thrilled the world with such harmonies as have swept into his consciousness when he has been alone with the seas and the stars? What philanthropist ever did all for the amelioration of man that he longed to do? What philosopher ever solved all his problems? What astronomer ever completed his explorations? Who ever truly loved without dreaming that love would spring a bridge over all abysses and follow its object through all spaces? Why are men afraid to die with guilty secrets unconfessed? Even the vilest and most hardened have dreaded the dreams that might come in the sleep of death. Aspiration, love, conscience, remorse, are as real as stones and stars, and if there is no possibility of their satisfaction each human being is a living lie. But suppose life does continue— what pictures Raphael may have made since he left the earth! What songs Milton, with his eyes open, may have sung! What music Mozart may have produced since the last wail of his Requiem sounded in his earthly ears! Then love may claim its own everywhere!

If there is no future life a singer is less than his song, an artist inferior to his picture, an architect not so enduring as the building which he designs; then Wagner's music draws thou-

sands across the sea to Baireuth while Wagner
himself has long ceased to exist; then Peter's
Dome will endure for countless generations but
Michael Angelo is but formless dust: then his
Dialogues will stimulate thought to the end of
time, but Plato for thousands of years has been
only a name; then the Teacher of Nazareth is as
dead as the cross upon which he was crucified. If such things are true, nothing can be
believed. But reason, conscience, aspiration,
love, rise in indignant protest and declare that
the thing made can not be greater than its
maker.

If the hypothesis of life beyond death is
unfounded, the wisest and best of all times have
believed in what has no reality. It can be
affirmed that no thinker of first‑rank whose
writings are preserved ever argued against immortality. Dr. Munger has well said: "The master-
minds have been strongest in their affirmations
of it. We do not refer to those who receive it
as a part of their religion. In weighing the
value of the natural or instinctive belief, Augus-
tine's faith does not count for so much as Cic-
ero's, and Plato's outweighs Bacon's: Plutarch is
a better witness than Chrysostom; Montesquieu
than Wesley; Franklin than Edwards; Emerson
than Channing; Greg's hope is more significant
than Bushnell's faith. All the great minds, often
in spite of apparently counter-philosophies, draw
near to the doctrine, and are eager to bear testi-

mony to it. Even John Stuart Mill, whose religious nature was nearly extirpated by an atheistic
education, does not say nay when the roll of the
great intellects is called. Blanco White, another
wanderer from the fold of faith, wrought into the
form of a sonnet so perfect that we instinctively
call it immortal, an argument, the force of which
men will feel so long as ' Hesperus leads the
starry host ':—

> ' If light can thus deceive,
> Wherefore not life ? '

Wordsworth touched the high-water mark of
the literature of the century in his ode on Immortality, and Tennyson's greatest poem is throughout exultant in the hope that ' Life shall live
forevermore.' " *

If there is no life beyond the grave, all moral
and religious thought, too, has been on a false
basis. The Greeks believed in immortality, and
reared exquisitely sculptured tombs to the
memory of their loved ones. In it the Egyptians
and Assyrians believed. It occupies a large place
in all religions. It may be a question whether
the Hindu Nirvana postulates continuance
of consciousness, but, while so many scholars
insist that it does, the Hindu faith cannot be
considered an exception. All men and nations
as far back as history goes have had some form
of religion. Their religion has been their most

---

* *Freedom of Faith*, p. 241.

sacred possession. It has embodied all that was
dear of hope, aspiration, faith ; and in the heart
of almost all forms of religion, from the baldest
paganism to the most exalted Christianity, has
been belief in life surviving death.

It may be said that no one of these arguments
is strong enough in itself to be conclusive, but
when all are woven into one we have a cumulative
argument which has the strength of certainty, a
chain which binds to the noblest life, the brightest
hopes, and the grandest achievement. That the
immortal life has not been demonstrated is most
freely granted, and that at present demonstration
seems to be forever impossible to those who are
in the flesh ; but we have made a credible basis
for faith : we have reached a point in which the
only thing for a reasonable being to do is to live
as if the immortal life had been mathematically
proven. To take any other course would be an
act of intellectual folly which would imply that
reason had ceased to have weight. I close this
part of the discussion with an impressive quota-
tion from a book which in recent days has helped
many to find a rational basis for faith in the
immortal life :

" From the first dawning of life we see all
things working together toward one mighty goal,
the evolution of the most exalted spiritual qualities which characterize humanity. Has all this
work been done for nothing? Is it all ephemeral,
all a bubble that bursts, a vision that fades? On

such a view, the riddle of the universe becomes a riddle without a meaning. The more thoroughly we comprehend that process of evolution by which things have come to be what they are, the more we are likely to feel that to deny the ever-lasting persistence of the spiritual element in man is to rob the whole process of its meaning. It goes far toward putting us to permanent intellectual confusion, and I do not see that any one has as yet alleged, or is ever likely to allege, a sufficient reason for our accepting so dire an alternative. For my own part, therefore, I believe in the immortality of the soul, not in the sense in which I accept the demonstrable truths of science, but as a supreme act of faith in the reasonableness of God's work." *

But what is thus prophesied is not mere existence: it is life. Existence alone is not desirable. What our Lord promised was not continuance of existence, but fullness of life. He said: "I am come that ye might have life, and more abundantly." Job would have derived no comfort from the idea of continued existence only, for that might have meant added misery. But because the future offered an opportunity of improvement, he hailed its suggestion with gladness. Existence tied to sorrow and living death is hateful; but that toward which the Apostle pointed was life which gives promise of infinite joy, for he

---

* John Fiske. *Destiny of Man*, pp. 113-116.

said: "This corruptible must put on incorrup-
tion: this mortal must put on immortality."
Life signifies health, growth, beauty, power, fruit-
fulness. In the Scriptures it never has any other
meaning. Spiritual life has one antithesis, and
that is death. It was not worth while for our
Master to come, if all that is before man is the
possibility of continuing to exist and think. If
life and existence are synonymous, the Buddhists
reason wisely when they say that there is nothing
so desirable as cessation of being, since existence
is the cause of sorrow. The word resurrection
carries with it a better significance—the rising of
one who has been down; newness of strength and
freshness of vigor. It is a hint of a glorious dawn
following a dismal night. The Greeks spoke of
the day rising from the bed of the night. The
teachings of our Lord are in harmony with the
prophecies of experience, and the voices that
sybil-like speak their oracles in the soul: and all
add emphasis to the words of the Apostle, "It is
sown a natural body, it is raised a spiritual body."
Life continues, with all its possibilities.

Life always moves toward blessing. If by any
chance it is deflected, and becomes ugly and evil,
it remains in that form but a little while, for its
tendency is to revert to its original type. Therefore we argue that if life continues it carries with
it by its very nature the possibility and necessity
of working itself clear from restrictions, limita-
tions, and corruptions. The experience of the

past, the philosophy of the ages, the profoundest poetry, the loftiest utterances of sage and seer, all accord with the simple and beautiful declaration of the Master, "If it were not so, I would have told you."

Life prophesies its own continuance; the history of thought utters the same prophecy; the religions of the world have added their witness— death is the freeing, not the ending, of life. But now the question rises, Have we any knowledge of what that life is? From that strange land for which all long have any returned, giving information of what it is like? Have human feet ever trod the slopes of those mountains, or human eyes gazed upon the spires of that city? The Bible is full of hints of the future, but contains little of revelation. We catch glimpses of glory, as light falls through rifts in the clouds on far-away heights, but that is all. Christ was more anxious that men should be in possession of the deathless life than able to satisfy curiosity concerning its nature. Nevertheless there are a few hints which may well be studied.

"In my Father's house are many mansions." As in a palace there are many rooms, so in the universe there are many spheres, and the children of God move at the Father's will from one to another. Dying, then, is not going into an abyss, but passing from one room to another in an infinite palace, and death is the door between the rooms.

The example of our Lord teaches us that resurrection is in a form, for the risen Jesus was known to Mary in the garden, to Thomas in the upper room, and to the disciples by the lake. He even used a human voice. He seemed to walk, as on the way to Emmaus. This teaches that heaven may not be beyond the stars, but in the very houses in which we live, and along the streets through which we move. Heaven may not be separated from us by distances, but by qualities: it may mean not so much a change of locality as of character. The appearance of the Master after death teaches that spirit is independent of physical barriers, for he appeared to Thomas in the upper room, and disappeared as if there were no physical obstructions in the walls.

After he had risen the Master continued his ministry. He talked with the disciples, gave them messages concerning his work, and to Peter he said: "Feed my sheep; feed my lambs." Jesus was the typical man. If he rose from the dead, we, too, shall rise : if he lives in a form, we, too, shall live in a form ; if his spiritual body is independent of physical barriers, we, too, shall be independent of them ; if after death he ministered to those who were left behind, then the spaces by which we are surrounded may be filled with ministering spirits. The heathen often caught dim glimpses of glorious truth. They spoke of "tutelary deities," whom they believed to be guardian spirits. The truth beneath their rude conception was clearly taught by our Lord when in ministry he returned to the earth. We may not be the only beings who walk these streets and gather in these churches. The appearances of Jesus after his resurrection teach that those who on earth had within them the slightest germ of the Divine Life, according as that life has grown, have followed in the footsteps of Him who solved the doubts of Thomas, spoke a loving word to poor bewildered Mary, and, after his denial, put upon Peter the burden of a blessed service.

That life in the spiritual body is free from sorrow and pain is taught in the seventh chapter of the Revelation, in words whose music sometimes seems the sweetest that ever broke upon human ears: "There shall be no more pain, and no more sorrow, and no sin, and no more death, and God shall wipe away all tears, and the Lamb shall be the light thereof."

But may it not be that humanity moves from the sorrow of the present into the joy of a future which is limited, and that *at last* death will be the end? We cannot enter that realm, but our Master has declared that life is everlasting. Moreover, it is the nature of all good things to endure : only evil tends toward death. If, now, we can think of life independent, and free from baleful influences and agencies, it is easy to believe that the word of our Lord was true when he spoke of the everlasting life. But there are

questions enough in the present without troubling
ourselves about the details of the future.   In the
meantime let us rejoice that the darkness of this
night breaks into the splendor of a new day, and
believe that no day will ever dawn which will close
in endless night.

If you ask me once more, why I believe all this,
I am not slow to reply :   I believe it because it is
reasonable,   because   any   other   hypothesis   is
unreasonable ; because it is an insult to human
powers to think that every voice which speaks
within our souls utters a long and persistent lie.
I believe it because it is right that we should take
counsel of the best within us rather than of the
worst.   It is more reasonable to rest in our hopes
than in our doubts, to have faith in things which
point toward blessing than to believe that there
is no purpose in creation, and that we, and the
universe, are driven hither and yon as drift on the
ocean, or leaves in a tornado.   I believe in the
immortal life because nothing else is reasonable.
There are a thousand questions which cannot be
answered.   In vain we reach for parallels in the
present, forgetting that the present can furnish
no   line   by   which   to   measure   the   future.   I
believe in the immortal life because nothing else
satisfies.   We are not like stars without orbits.
We are not made to sing songs that will make
music after we are nothing.   We are not here to
build houses which endure while we dissolve in
dust.   We are not here to love, aspire, grow

strong and noble, and then see that it would have been just as well to have reached after nothing, to have hated as to have loved, because all things move into one bottomless abyss. I believe in the immortal life because it is the noblest and most inspiring view of the life which now is. It makes the best and happiest men, and it will make the best and happiest world. That cannot be altogether unreal which moves toward the highest happiness and the supreme welfare. Only grant the future, and all the sorrows of time may be righted, all the inequalities and injustices of the present may prove to be ministers of blessing rather than of cursing. That which makes the best world is the best faith, call it what you will. I believe in the immortal life because the noblest Being who ever walked this earth, whose whole career was instinct with truth and love, who proved his right to speak for God by his likeness to God, has declared that we are in our Father's house, that what we call death is only sleep, and that those who rest in that sleep will in the morning awaken in a brighter room to sweeter service.

If, then, after all we are mistaken ; if the great and the good in the past—the greatest and best, let us say—have believed a lie ; if the voices which speak within our own souls utter a lie, and we are mistaken and death ends all—what then? Well, there is only one answer : even if it be error, it is a blissful one. Better to live sweet

and noble lives with so blessed an illusion than
to grope aimlessly in the darkness, without light
or hope. But we are not mistaken. The faint
intimation which flashed upon the wondering
vision of Job has grown into full-orbed light as it
falls upon our faith and love. We have heard
His voice speaking to us who said, "I am the
resurrection and the life." We have seen men,
buried in the grave of sin, rise in newness of life
on the earth. We have seen those whom we
loved go down into the dark valley with songs
upon their lips. We are seeing in our time the
curtain that separates the visible from the
invisible grow thinner and thinner, until no one
would be surprised if it should be rent in twain.
We will not believe that life, history, human
thought, all the world's religions, all the world's
bibles, and our Master Jesus Christ, are all
deceivers, and that a falsehood is at the base of
all things. We will rather trust, hope, love,
aspire, and work on toward that which is best,
assured that by and by the day will dawn and the
shadows flee away; that we shall see Him and be
like him; that the light affliction which is but for
a moment will work a far more exceeding and
eternal weight of glory. We will keep near to
Him through life, and through death, until with
Him we realize the fullness of his words—*Resurrection, and Life !*

www.ingramcontent.com/pod-product-compliance
Lightning Source LLC
Chambersburg PA
CBHW021416090426
42742CB00009B/1162